Perilous Charmers

Poisonous Plants of the Pacific Northwest

By Mary Tapson-Jones

M L Tapson-Jones

Edited by Graham Boffey

Published by Fran Kay & Associates
Chase, British Columbia, Canada

Canadian Cataloguing in Publication Data
Tapson-Jones, Mary, (date)

Includes bibliographical references and index.
ISBN 0-9698250-3-X

1. Poisonous plants --Northwest, Pacific. I. Title.
QK100.N78T36 1995 581.6'9'09795 C95-911120-4

Photos by Mary Tapson-Jones
Edited by Graham Boffey
Printed in Canada by New Horizon Printing, Summerland B.C.

Distributed by Fran Kay & Associates
R.R.#2, Chase B.C. Canada V0E 1M0
Phone/Fax 604-679-5362
E-mail: frankay@netshop.net

Acknowledgements

Salmon Arm, B.C. 1995

People today do not have parents or grandparents with the knowledge of generations of tried and true use of wild and domestic plants. Thus, these people of the 20th and 21st centuries are susceptible to poisoning themselves or others through use of unidentified and unknown plants in the preparation of teas and salves. The following includes most of the known dangerous wild and domestic plants found west of the Rocky Mountains from Oregon north.

I wish to thank Dr. Barry Lipsett for starting and encouraging me to work on this project. Likewise, I cannot thank Marg Shand and all my other friends enough for their help and support over these many years. And without my husband's earlier encouragement and belief in my ability, I would never have attempted this project. My background comes from generations of gardeners and botanists, thus my keen interest in tracing the origin and use of plants of all kinds. I learned plant names and terminology from the time I could walk, thanks to my mother.

I also thank Graham Boffey, toxicologist at the Royal Inland Hospital in Kamloops for his kind assistance in checking the entries, his additions, editing, constructive comments and for writing the Forward. And Dr. Don Paterson for his interest and good advice.

I wish to dedicate this book to my mother and to my husband.

M.L.Tapson-Jones

Forward

Growing interest in family outdoor recreation and the revival of the use of folk herbal remedies for the treatment of minor ills has increased the risk of poisoning by plant ingestion and contact. This is particularly apparent in British Columbia where there has been a rapid increase in population growth during recent years, most of it urban, but with many of it's new residents partaking in outdoor pursuits such as hiking and camping for the first time. Each year brings it's admissions to hospital emergency departments of patients who have ingested plant materials of various kinds. These patients are often children. Fortunately, most of these cases have a favorable outcome, but the possibility remains that a serious accident could occur, as many plant species in the Pacific Northwest have the potential to poison an unsuspecting individual.

This book is a timely publication, in which the author provides useful information in a clear, concise manner, assisted by photographs which aid in the recognition of poisonous plant species, facts on the identity of toxic agents, and their effects on humans and livestock.

Graham Boffey
Toxicologist, Royal Inland Hospital, Kamloops B.C.

Table of Contents

Legend

Red	—	Most toxic, obtain medical attention IMMEDIATELY.
Yellow	—	Toxic, obtain medical attention soon.
Blue	—	Skin irritations and non-life threatening.

Wild Plants

WATER HEMLOCK is a member of the carrot family (Umbelliferae) and is found beside, and in, streams all over North America. It grows 1 to 2 1/2 metres (3 to 7 ft.) tall, with large alternate, divided, lanceolate leaves and leaf stalks which clasp the stem. The central vein runs the length of the leaflet, while secondary veins run into the valleys or notches between the points on the edge of the leaves. The umbels of tiny creamy-white flowers closely resemble carrot flowers but grow further above the leaves. The roots are a fleshy bundle, unlike the single taproot of the non-toxic wild parsnip. Because water hemlock is easily pulled up and eaten with the grass, many animals have died a very painful death. A piece of stem the size of a walnut is fatal to a human, and very little more is required for a bovine. The poison acts so fast that casualties have been found beside the water with parts of the plant still in their mouths. <u>This species is considered the most poisonous of the temperate zone.</u> **Common Name:** Water hemlock, beaver poison, cowbane. **Botanical Name:** Cicuta occidentalis and similar species. **Toxicity:** All parts. Active component is contained in the yellowish, oily liquid in the plant tissues, known as cicutotoxin, a highly unsaturated higher alcohol. **Signs & Symptoms:** Acts directly on the central nervous system. Symptoms usually appear within half an hour of ingestion of a lethal dose, and can range from 15 minutes to more than an hour. Initially, excessive salivation and nausea occurs, followed by tremors and extremely violent convulsions. Death is a result of respiratory failure. If vomiting has occurred as one of the earliest symptoms, chances of survival are significantly improved. **Treatment:** EXTREMELY POISONOUS, OBTAIN MEDICAL ASSISTANCE IMMEDIATELY.

Water hemlock

CHOKECHERRY is a shrub which grows from 1 to 3 1/2 metres (3 to 13 feet) tall. The flowers and fruit hang in pretty compact racemes. Bitter cherry is a small tree with reddish bark often 6 1/2 to 10 metres (20 to 30 feet) tall under good growing conditions. The flowers grow in small clusters all over the tree. Though the flesh of most cherries is edible, all their seeds or stones are dangerous when ingested because of the presence of a cyanogenetic glycoside which is readily hydrolyzed to hydrocyanic acid. The leaves and bark contain the same poison; therefore ingestion of teas brewed from these materials is not recommended. **Common Names:** Chokecherry, western chokecherry, pincherry, bitter cherry etc. **Botanical Name:** Prunus demissa, Prunus pensylvanica and related species. **Toxicity:** All parts except the flesh of the fruit, due to hydrolysis of the cyanogenetic glycoside, amygdalin. **Signs & Symptoms:** Early stimulation of respiration, changing to gasping, staggering, paralysis, convulsions and death in extreme cases of poisoning. **Treatment:** OBTAIN MEDICAL ASSISTANCE IMMEDIATELY.

BANEBERRY is a robust many branched plant that grows 1 metre (2 to 3 ft.) tall in cool damp woods, often near streams. The sharply toothed, compound green leaves stand out clearly from all the other vegetation. In June, the racemes of creamy white flowers appear to float above the leaves like dainty bottle brushes, while in late July and August, the shiny bright, light to dark red and sometimes white berries tempt the unwary to eat them. When the leaves have fallen, these berries remain untouched by all creatures of the forests, even insects, thus warning humans not to touch them. Ingestion of as few as six berries can cause severe symptoms in humans. **Common Names**: Baneberry, dolls-eyes, snakeberry, white cohosh, coralberry. **Botanical Name:** Actaea rubra, Actaea pachypoda. **Toxicity:** All parts are poisonous and are thought to contain a toxic essential oil or glycoside. **Signs & Symptoms:** Acute stomach cramps, headache, gastroenteritis, vomiting, circulatory collapse. **Treatment:** OBTAIN MEDICAL ASSISTANCE IMMEDIATELY.

Chokecherry

Baneberry

DEATH CAMAS belongs to the lily family and in the early spring it's bright green leaves appear amongst the early grass spears and present a hazard to grazing animals who may ingest the plant along with the grass. One mouthful of the leaves of the Death Camas can be fatal. The bulb was occasionally dug in mistake for the wild onion by Indians and earlier settlers when digging for wild bulbs and corms. Only one or two of these little bulbs were needed to make the stew deadly. However, considerable variation in toxicity between the various species exists. **Common Name:** Death camas. **Botanical Name:** Zigadenus venenosus and related species. **Toxicity:** All parts are poisonous and contain steroid alkaloids such as glycoalkaloids and ester alkaloids including zygadenine and zygacine. **Signs & Symptoms:** Muscular weakness, gastroenteritis, vomiting, diarrhea. **Treatment:** OBTAIN MEDICAL ASSISTANCE IMMEDIATELY.

MOUNTAIN MARSH MARIGOLD forms a clump of bright green, heart shaped leaves that grow in marshy areas and by mountain streams. The leaves remain green all winter under the snow and are often eaten by elk, though few other creatures can eat this bitter plant with impunity. The flowers are white with a bluish tinge on the backs of the sepals. Marsh Marigold has yellow flowers and sepals but is otherwise similar to the white variety. **Common Name:** Mountain marsh marigold, marsh marigold. **Botanical Name:** Caltha leptosepala, Caltha palustris. **Toxicity:** All parts are poisonous, due to alkaloids including helleborin and anemonin. **Signs & Symptoms:** May cause blistering of the mouth, gastric distress, abdominal pain and diarrhea. Skin contact with the juice may cause reddening, dermatitis and blistering. **Treatment:** OBTAIN MEDICAL ASSISTANCE IMMEDIATELY.

Death camas

Marsh marigold

POISON HEMLOCK is an introduced biennial with dark green, parsley-like foliage and is also a member of the carrot family (Umbelliferae) like the water hemlock. It may be found anywhere it can find sufficient soil to grow. The young, first year leaves have been mistaken for parsley, with dire results. In it's second year, a many branched stem with purple spots develops, growing to 2 to 2 1/2 metres (6 to 8 feet). The small white flowers grow in a broad umbel, well above the main structure of the plant. The seeds look similar to those of dill, parsley and anise, and are toxic. Centuries ago, in Europe, a tea made of this plant was given to condemned criminals as a form of execution. **Common Name:** Poison hemlock, poison parsley, spotted hemlock. **Botanical Name:** Conium maculatum. **Toxicity:** All parts. At least five distinct alkaloids are contained in the tissues of this plant, all of which have close similarities to nicotine. The most well known is the volatile, colorless, strongly alkaline oil, conine. **Signs & Symptoms:** Nervousness, tremor, staggering gait, dilation of the pupils, slowness of the heartbeat, coma and death by respiratory paralysis. Symptoms may be observed as early as 12 minutes after ingestion. **Treatment:** OBTAIN MEDICAL ASSISTANCE IMMEDIATELY.

RED ELDERBERRY is a fast growing shrub with pithy, brittle stems. The flowers grow in large, terminal triangular-shaped clusters 5 to 8 cms. (2 to 3 in.) across. The leaves are opposite and pinnately divided, consisting of 5 to 11 lanceolate, toothed leaflets per leaf. The bright red berries cause nausea if eaten raw but are edible when cooked. Ingestion of the roots and use of the hollow stems as whistles or blow guns by children has resulted in poisoning, though adults are rarely affected. Thus care must be taken not to endanger children's lives by making toys of these stems. **Common Name:** Red elderberry. **Botanical Name:** Sambacus pubens and related species. **Toxicity:** Leaves, stems and roots contain a poisonous alkaloid and a cyanogenic glycoside. **Signs & Symptoms:** Diarrhea, nausea. **Treatment:** OBTAIN MEDICAL ASSISTANCE IMMEDIATELY.

Poison hemlock

Red elderberry

MOUNTAIN RHODODENDRON is found in the damp under growth and on avalanche slopes from about 900 metres (3000 ft.) to treeline on the western mountains of North America. It is an erect, deciduous shrub which has earned uncomplimentary names because of the difficulty walking through dense stands of it. The leaves are thin and elliptical from 3 1/2 to 7 1/2 cms. long and the creamy-white flowers occur singly or in groups of up to three. Like all relatives of the laurel, members of the rhododendron family are poisonous. Poisoning has resulted from ingestion of a tea brewed from the leaves of this and related species. **Common Name:** Mountain rhododendron, mountain misery, tangle-bush. **Botanical Name:** Rhododendron albiflorum and related species. **Toxicity:** All parts contain the resinoid carbohydrate, andrometoxin. **Signs & Symptoms:** Time from dosage to first appearance of symptoms is variable and evidenced by excessive salivation, watering of the eyes and nose, slow pulse, vomiting, convulsions and progressive paralysis of the arms and legs, then death. **Treatment:** OBTAIN MEDICAL ASSISTANCE IMMEDIATELY.

CORYDALIS has grey-green, many times dissected leaves, and is often found growing on gravel banks or in forest areas that were burnt over. The yellow flowers grow in racemes on 6 to 12 cm. (3 to 6 in.) stems, much like some of the vetch family, or the bleeding heart. If crushed, the foliage and root have a very strong unpleasant odor which should warn anyone to leave it alone, but it has been eaten, with dire consequences. **Common Name:** Corydalis. **Botanical Name:** Corydalis and related species. **Toxicity:** All parts contain corycavine. **Signs & Symptoms:** Trembling, convulsions. **Treatment:** Treat like an overdose of narcotic. OBTAIN MEDICAL ASSISTANCE IMMEDIATELY.

Mountain rhododendron

Corydalis

FALSE HELLEBORE applies to two species of Veratrum. Veratrum viride grows in low, moist habitat across North America, while Veratrum californicum prefers alpine meadows in the Pacific northwest, sometimes growing as tall as 2 metres (7 ft.). It's pronounced parallel-veined leaves clasp the stem while greenish-yellow flowers stand gracefully erect or droop out from the leaf bract. Though not eaten by many animals, bears use it as a spring purgative. Horses that have eaten it can often be saved by walking them until the effect of the relaxant wears off. Native peoples dipped hunting arrows in false hellebore. Once the poison was in the blood stream, the animal would become drowsy, lie down to sleep and not awaken. Today it is used in some high blood pressure medicines. The powdered rhizome is used as an insecticide. **Common Name:** False or white hellebore, Indian poke. **Botanical Name:** Veratrum viride, Veratrum californican. **Toxicity:** All parts contain glycoalkaloids and also ester alkaloids such as veratrosine and veratrin. **Signs & Symptoms:** Salivation, nausea, diarrhea, gastroenteritis, slowing of the heart rate, breathlessness, headache. **Treatment:** OBTAIN MEDICAL ASSISTANCE IMMEDIATELY.

MOUNTAIN LAUREL can be found in damp places from elevations of 450 metres (1500 ft) to high mountain meadows. The leaves of this tiny shrub are ovate, leathery and evergreen with white fuzz on the undersides, approximately 2 1/2 to 6 cms. long, mostly opposite on upright branches. Flowers are rose colored and saucer-shaped. The plant grows in many places where sheep are pastured, so shepherds must watch the lambs carefully; ingestion of a single leaf can be fatal. The larger species of this shrub might be mistaken for Labrador tea and one can only assume that this has contributed to the incidence of fatalities in humans. **Common Name:** Mountain laurel, lambkill, sheepkill, calfkill, dwarf laurel, wicky. **Botanical Name:** Kalmia augustifolia, Kalmia carolina, Kalmia polifolia. **Toxicity:** Leaves, stems, flowers and pollen contain a resinoid carbohydrate, andromedotoxin. **Signs & Symptoms:** Salivation, watery eyes and nose, slow pulse, vomiting, convulsions, progressive paralysis, death. **Treatment:** OBTAIN MEDICAL ASSISTANCE IMMEDIATELY.

False hellebore

Mountain laurel

WILD RHODODENDRON is quite a large, coarse-branched, evergreen shrub that grows in woods and thickets in limited areas of British Columbia and south into the Washington and Oregon states. It grows from 1 to 5 m. (3 to 16 ft.) tall and bears long, elliptical, leathery leaves mostly at the tips of the twisted branches, as do the many showy white to pink flowers. The rhododendron is a member of the laurel family and like all members of this family, is poisonous. **Common Name:** Wild rhododendron, Californian rose bay. **Botanical Name:** Rhododendron macrophyllum. **Toxicity:** All parts contain a resinoid, glycotoxin, andromedotoxin. **Signs & Symptoms:** Salivation, slow pulse, low blood pressure, vomiting, convulsions, paralysis, death. **Treatment:** OBTAIN MEDICAL ASSISTANCE IMMEDIATELY.

MONKSHOOD is a shy perennial herb with erect or slightly trailing stems growing from 30 cms. to 2 m. (2 to 6 ft.) tall. The leaves are alternate, dark green, roundish, deeply palmated with 3 to 9 lobes very similar in appearance to those of a buttercup. The dark blue-purple flowers are a terminal raceme shaped like the hood or cowl worn by monks. The petals form the lower, boat-like cup underneath the cowl, making the flower strikingly unusual and a treat to find in woodland glades over much of North America. Several species have been cultivated as garden plants. **Common Name:** Monkshood, wolfsbane, western monkshood, aconite. **Botanical Name:** Aconitum columbianum. **Toxicity:** All parts contain the polycyclic diterpenoid alkaloid, aconitine and other alkaloids. **Signs & Symptoms:** Restlessness, salivation, weakness, tachycardia, nausea, vomiting, diarrhea, respiratory paralysis, convulsions and death. **Treatment:** OBTAIN MEDICAL ASSISTANCE IMMEDIATELY.

Wild rhododendron

Monkshood

YELLOW FLAG was introduced from Europe as a garden plant. This member of the iris family is now found growing in moist, marshy areas in many parts of North America, including parts of British Columbia. The plant has long, erect, linear-veined leaves attached to thick, fleshy rootstalks. All members of the iris family are poisonous, especially the roots or rhizomes. Animals and children have been known to eat the young green growth when it first comes up in the spring, or to have nibbled on the soft roots. **Common Name:** Iris, yellow flag, or fleur-de-lis. **Botanical Name:** Iris pseudacorus. **Toxicity:** All parts are toxic, particularly the rhizomes, which contain the emetic and cathartic, irisin. **Signs & Symptoms:** Gastroenteritis, diarrhea, vomiting, coma and death. **Treatment:** OBTAIN MEDICAL ASSISTANCE IMMEDIATELY.

WILD LARKSPUR often grows 15 to 180 cms. (6 to 75 in.) tall. Flowers are blue, sometimes pink or white, bilaterally symmetrical in erect terminal racemes, the upper sepal elongated backwards to form a spur. Larkspurs are perennial, erect herbs with leaves that are simple, alternate, or clustered at ground level. In some, the lower ones are palm-like, with small spikes of flowers above. Sheep and cattle may graze them readily with young grass. Children have been poisoned by the small pea-like seeds. Western larkspurs occur as tall (Delphinium occidentale) and low (Delphinium nelsoni) varieties. Tall,1 to 2 1/2 metres (3 to 7 ft.) varieties like high mountain elevations; low ones are found in moist locations in the foothills. Larkspurs belong to the Ranunculaceae family. **Common Name:** Larkspur, delphinium. **Botanical Name:** Delphinium occidentale, Delphinium nelsoni. **Toxicity:** All parts contain diterpinoid alkaloids such as delphinine. **Signs & Symptoms:** Gastroenteritis, nausea, constipation, respiratory paralysis and death. **Treatment:** OBTAIN MEDICAL ASSISTANCE IMMEDIATELY.

Yellow flag

Wild larkspur

LUPINS have pea like flowers ranging in color from dark blue to white, and occasionally pink. Members of this family can be found growing from sea level to alpine meadows. Some species have their palm-like, gray-green leaves as close to the ground as 5 cm. (2 in.) with small spikes of flowers above them; other species may grow as tall as 50 cms. (20 in.) above the roots. Most species are perennial and the fruit is the characteristic flattened pod of the Leguminosae, containing several pea-like seeds, deadly to children. Cattle and sheep may graze it or be fed it in the hay. **Common Name:** Lupin, blue bonnet, silky lupin. **Botanical Name:** Lupinus sericeus, Lupinus leucopsis, Lupinous laxiflorus. **Toxicity:** All parts, particularly the seeds, contain lupine alkaloids such as lupinine. Not all are poisonous. Minor changes in chemical structure appear to cause major changes in toxicity. **Signs & Symptoms:** Labored breathing, sometimes spasms, coma and death from respiratory paralysis. **Treatment:** OBTAIN MEDICAL ASSISTANCE IMMEDIATELY.

WILD BLEEDING-HEART, with it's soft, many times dissected blue-green leaves, grow in damp, cool places, mostly near the coast, in wooded areas, along hedgerows and in clearings. In the southern interior of British Columbia, the domesticated plant has escaped from gardens and become established in the wild. It flowers in early spring and produces pendant, bilaterally symmetrical, showy blooms in which the petals are continued backward into two spurs. Colors vary with the species; in the west, they are rose-purple. All members of this family are poisonous. DO NOT use for food any plant with foliage that resembles parsley. Some of our most poisonous plants have leaves that look like some of our garden herbs. **Common Name:** Bleeding heart, steershead, staggerweed, Dutchman's breeches. **Botanical Name:** Dicentra formosa, Dicentra spectabilis. **Toxicity:** All parts, including the roots, contain isoquinoline alkaloids - aporphine, protopine, protoberberine, except for Dicentra spectabilis, from which only protopine has been isolated. **Signs & Symptoms:** Trembling, salivation, projectile vomiting, convulsions. **Treatment:** OBTAIN MEDICAL ATTENTION IMMEDIATELY.

Lupin

Wild bleeding heart

The erect, many-branched stems of **ST. JOHNSWORT** grow as tall as 60 cms. (24 in.) and occasionally to a metre (39 in.) Linear-oblong leaves are opposite, attached directly to the main stem, and spotted with dark colored dots. The bright yellow flowers have numerous long, attractive stamens which are one of the distinguishing features of this species. Several are native to western Canada, but the most dangerous one is an immigrant from Europe. When any part of this plant is eaten, or consumed as tea for a period of time, primary photosensitization to sunlight occurs, the first plant recorded with this characteristic. **Common Names:** St. Johnswort, goatweed. **Botanical Name:** Hypericum perforatum, Hypericum formosun, Hypericum majus. **Toxicity:** All parts contain a photodynamic, naphthodianthrone pigment known as hypericin. **Signs & Symptoms:** Intense itching, red skin, fluid-filled swellings under the skin and/or leakage. Dead skin sloughing off. Rarely fatal but death following convulsions has been reported. In livestock, it may cause inflammation of mucous membranes, subsequent refusal to eat, and starvation. **Treatment:** Protect from sun. OBTAIN MEDICAL ASSISTANCE IMMEDIATELY.

YEW is the only evergreen conifer growing in western North America that produces berry-like seeds. The pulp of this 'berry' is not poisonous, but the seed is and should never be eaten. Most parts of this tree are poisonous; therefore care must be taken not to prepare tea from any conifer unless the identity is certain. The western yew is not a large tree, nor is it very common. It can occasionally reach a height of 15 m. (45 ft.) and is usually found under the canopy of larger evergreens from California to Alaska. It's bark is dark red; it's needles dark green, narrow, stiff, and between .5 and 4 cms. long. **Common Name:**Yew, western yew. **Botanical Name:** Taxus brevifolia. **Toxicity:** Most parts, except for the flesh of the "berries", contain the alkaloid taxine. **Signs & Symptoms:** Diarrhea, vomiting, dilation of the pupils, muscular weakness, convulsions and death. **Treatment:** OBTAIN MEDICAL ASSISTANCE IMMEDIATELY.

St. Johnswort

Yew berries

BITTERSWEET is a woody vine, originally from Europe, that escaped from gardens and can be found growing wild in much of western North America. The dark green foliage is similar in shape to a bean leaf. May through July, it's yellow anthered, purple-petaled flowers hang from the stems, which can reach more than 2 m. (6 ft.) in length. The flowers are followed by bright red berries that hang in tempting clusters, often throughout the winter. Though all parts of this vine are poisonous, the berries are the most dangerous. Several cases of poisoning have been reported. **Common Name:** Bittersweet, European bittersweet, woody nightshade, climbing nightshade. **Botanical Name:** Solanum dulcamara. **Toxicity:** All parts contain glycoalkaloids, including solanine. **Signs & Symptoms:** Intact solanine irritates the gastrointestinal tract, causing stomach cramps, diarrhea (or constipation) and abdominal pain. Hydrolysis of the glycoalkaloids yields sugars and steroidal alkamines, which are more readily absorbed and affect the nervous system. This results in apathy, drowsiness, salivation, weakness, paralysis, coma and eventual death. **Treatment:** OBTAIN MEDICAL ASSISTANCE IMMEDIATELY.

Wild **SWEET PEA,** not to be confused with the bushy Hedysarum mackenzii, which is also known by this common name, is a vigorous, escaped garden vine. In northwestern North America, it is a common, persistent perennial with winged stems 1 to 2 m. (3 to 6 ft.) long. The flowers occur in rose and purple racemes. The fruit is a legume pod containing many seeds. All of the plant is toxic to livestock and humans, and responsible for many deaths. **Common Name:** Wild pea. Wild sweet pea. **Botanical Name:** Lathyrus sylvestris. **Toxicity:** All parts, particularly the seeds, contain at least two different toxic substances. Beta-(gamma-L-glutamyl)-aminipropionitrile, produces paralysis and skeletal lesions. Nervous symptoms result from the action of L-alpha, gamma diaminobutyric acid. **Signs & Symptoms:** If the peas make up 1/3 to 1/2 the total diet, these may occur - hyperexcitability, spasticity of leg muscles with the characteristic feet turned inward and toes down, paralysis, convulsions, death. **Treatment:** OBTAIN MEDICAL ATTENTION.

Bittersweet

Sweet pea

WILD SUMAC, like poison ivy, grows on warm sunny banks, often near lakes, and can grow to a height of 2 1/2 to 4 m. (8 to 12 ft.) The large compound leaves turn a bright red in autumn but at other times of the year it can easily be passed by, unnoticed. In winter, the large clusters of flower buds have a covering of red hairs, though the flowers them selves are green when open. The hairs that cover the shrub can cause much the same kind of rash as poison ivy, because they contain a similar oily irritant. Pollen from the blooming shrub is enough to cause a severe rash in sensitive individuals in close proximity to it. Do not confuse this plant with poison sumac, which is an eastern variety. **Common Name:** Sumac, Stag horn. **Botanical Name:** Rhus and related species. **Toxicity:** All parts, including pollen. **Signs & Symptoms:** Red itchy skin, blistering and weeping. **Treatment:** Wash as soon as possible with strong soap. OBTAIN MEDICAL ATTENTION AS SOON AS POSSIBLE.

POISON-IVY is a small, woody, climbing vine or non-climbing shrub, 15 to 30 cms. (6 to 12 in.) tall, that may be found growing in warm stony ground, along lakeshores, beside roads and fences. The leaves are alternate, with three leaflets, the small flowers greenish-white. The fruits are tan to yellow and hang in clusters. Care should be taken when walking where this species is growing, to avoid personal contact or that of pets with the plant. The oil from the leaves and broken stems readily transfers to clothes, footwear, animal fur, and on to the skin, where it can cause a very nasty, weeping rash. Burning the plant produces smoke containing droplets of the toxin and particles of ash coated with it. Exposure to the smoke can also produce a rash. **Poison oak** is a non-climbing shrub related to poison ivy, which it much resembles in appearance and effect. **Common Name:** Poison-ivy, Western poison-oak. **Botanical Name:** Rhus radicans or Toxicodendron radicans, Rhus diversiloba or Toxicodendron diversiloba. **Toxicity:** All parts contain catechol derivatives, including 3-n-pentadecylcatechol. **Signs & Symptoms:** Red itchy skin, blistering and weeping. **Treatment:** Wash with strong soap to remove oil. OBTAIN MEDICAL ATTENTION AS SOON AS POSSIBLE.

Wild sumac

Poison ivy

BUTTERCUPS usually have shiny yellow petals, but a few have white and some, almost none at all. Members of the Ranunculaceae family, they grow from sea level to high alpine areas, in freshwater, marshy areas, and some on quite dry land. Height varies from 2 1/2 cms. (barely an inch) to 50 cms. (25 in.) The stem leaves are alternate, palmate, and may be entire, lobed or dissected. One of the earliest of spring blooms, on warm, grassy meadows and rocky sidehills. All members of this family are poisonous. Animals usually avoid them because of the disagreeable taste but occasional poisoning still occurs, as do addictions. The sap of some varieties may cause skin blistering and ulceration. **Common Name:** Buttercup, crowfoot, water crowfoot, tall field buttercup. **Botanical Name:** Ranunculus and related species. **Toxicity:** All parts contain the glycoside ranunculin which breaks down by enzyme action into toxic protoanemonin, not a highly toxic substance. Its content in plants varies widely. **Signs & Symptoms:** Irritation of the oral tissues, salivation, diarrhea and abdominal pain. **Treatment:** If ingested, OBTAIN MEDICAL ASSISTANCE.

HORSETAIL represents a botanical family over 400 million years old, when it grew to more than 25 m. (80 ft.) Those in the Pacific northwest today grow only 25 cms. to 1 1/2 m. (10 to 50 in.) tall. These are herbaceous perennials with simple or branching erect stems, hollow except at the nodes. They reproduce by releasing spores from a cone-like sporangia at the end of separate fertile shoots. Aboriginal people and early settlers in the Northwest once used the abrasive stems as wood polishers and pot scrubbers and in folk medicine, externally as a poultice, and internally as a diuretic and to reduce intestinal bleeding. **Common Name:** Horsetail, scouring rush. **Botanical Name:** Equisetum arvense, Equiset telmateia, Equiset hiemale. **Toxicity:** All parts contain thiaminase, an enzyme which destroys the vitamin thiamine. Large variations in the degree of toxicity occurs between the various species of Equisetum and the susceptibility of various animal species. **Signs & Symptoms:** Hyperexcitability, weakness, ataxia, diarrhea and occasionally, death. **Treatment:** OBTAIN MEDICAL ASSISTANCE.

Buttercup

Horsetail

GROUNDSELS vary from 15 cms. (5 in.) for some coarse, bushy perennial varieties to 1 1/2 m. (50 in.) for tall, erect annuals. The small daisy-like flowers are usually yellow and consist of a single whorl of bracts which touch but do not overlap, arranged on top of the stems. Some have few or no rays. Found worldwide. Not all groundsels are known to be toxic. Those that are appear to contain the highest concentration in the seeds. Human deaths have occurred in Africa when flour has been contaminated with groundsel seeds. Livestock losses are significant. Poisoning is insidious because it affects the liver, and may take weeks or months. **Common Name:** Groundsel. **Botanical Name:** Senecio vulagaris, Senecio integerrimus. **Toxicity:** All parts contain pyrrolizidine alkaloids such as retrorsine. **Signs & Symptoms:** Accumulation of fluid in body cavities, enlarged liver, abdominal pain, nausea, vomiting, diarrhea, blood blisters, coma, death. In livestock, usually fatal within a week. Yellowish discoloration of mucous membranes, symptoms similar to humans. **Treatment:** OBTAIN MEDICAL ASSISTANCE.

CLEMATIS is a native vine that can grow from 1 to 10 m. (3 to 30 ft.), climbing up and over anything in it's path. The blueish flowered variety grows in cool shady places, hanging over shrubs and fences, while it's white-flowered sister grows on hot, rocky slopes. The fluffy seed heads give this vine one of it's common names. The species has a reputation as being poisonous, but no cases of livestock poisoning have been authenticated in North America. Clematis sap is an irritant, causing blisters. A clematis liniment was a popular treatment for itch and skin diseases in pioneer days. **Common Name:** Clematis, virgin's bower, old man's beard. **Botanical Name:** Clematis virginiana. **Toxicity:** All parts contain a vesicant (causing blisters) sap in which various alkaloids, glycosides and sapononins have been reported to occur. It is a distinct possibility, as a member of the Ranunculaceae family, the Clematis species contain protoanemonin or similar compounds. **Signs & Symptoms:** After ingestion, burning sensation of mouth, salivation, in extreme cases, bloody vomiting, diarrhea and convulsions. Contact of the sap with the skin may produce blistering. **Treatment:** OBTAIN MEDICAL ASSISTANCE.

Groundsel

Clematis

LOCOWEED can be either of two species, Astragalus or Oxytropis, and can be found in a variety of habitats ranging from warm, rocky places and dry, grassy pastures to moist forest and stream bank soils where domestic animals are apt to graze in early summer. The two species are very similar, differentiated by botanists by the shape of the flower keel. Astragalus has a short, blunt keel; Oxytropis' is long, with a distinct point. Low growing varieties can be mistaken for sweet clover. Animals may become addicted, refuse to eat anything else, and eventually die. Humans are reported to have made and drunk locoweed tea for many weeks and been affected the same way. The toxin affects sensory and motor nerves; thus the animal staggers or acts 'loco', the Spanish word for 'crazy'. **Common Name:** Locoweed, crazyweed. **Botanical Names:** Oxytropis, Astragalus species. **Toxicity:** All parts contain the toxic agent, the identity of which is not completely known. It does have some alkaloidal properties and is known as locoine. **Signs & Symptoms:** Staggering gait, difficulty eating and drinking, listlessness, paralysis, and in extreme cases, coma and death. **Treatment:** OBTAIN MEDICAL ASSISTANCE.

SPREADING DOGBANE is an erect, perennial herb with opposite-paired, smooth-margined leaves. The stem is fibrous and carries pretty pink to cream-colored flowers in loose clusters. The sap is milky and the plant is often wrongly called 'milkweed', a name which should be reserved for Asclepias species of the family Asclepiadaceae. There are two similar varieties of dogbanes which are common weeds of open ground and roadsides in most of North America. All members of this family have fibrous stems, once the most important source of plant fibre available to aboriginal people in southern British Columbia. This strong fibre has been used for centuries to make rope, weave mats and make fish lines. **Common Name:** Spreading dogbane, Indian hemp. **Botanical Name:** Apocynum cannabinum and Apocynum androsaemifolium. **Toxicity:** All parts are poisonous and contain apocynamarin. **Signs & Symptoms:** Gastric disturbances including nausea, diarrhea. **Treatment:** OBTAIN MEDICAL ASSISTANCE.

Locoweed

Spreading dogbane

WESTERN ANEMONE grows in damp alpine meadows and blooms very early in the season, pushing through snow banks as early as March at lower elevations. The flowers are cup-shaped, creamy white, purple or violet, and open well above finely cut leaves that are covered with fine, silky hairs, giving them the appearance of having a fur coat. When in seed, the flower stalk seems to sport a white tam-a-shanter, likened by some to armies of tiny men marching across the meadow. The anemone belongs to the buttercup family, the Ranunculaceae, and as such, is toxic and thus not eaten by many animals. **Common Name:** Western anemone, chalice cup, windflower, pasqueflower, thimbleweed. **Botanical Name:** Anemone patens and related species. **Toxicity:** Leaves and stems contain ranunculin which breaks down to produce toxic protoanemonin. **Signs & Symptoms:** Skin irritation and sometimes blistering. **Treatment:** Wash as soon as possible after exposure.

Western anemone

Garden Plants

AUTUMN CROCUS blooms in the fall, but the large dark-green leaves appear in spring and die down in the fall, as the plant awaits the cool autumn rains. The large, crocus-shaped mauve to pink colored flowers seem to appear from nowhere, thrusting up through the last of the summer annuals. The bulb, with it's stored autumn blooms, makes an attractive indoor plant when the bulb is placed in water and allowed to flower. It can then be planted back into the garden after flowering. Fatalities have occurred as a consequence of eating parts of the plant, including the leaves, because it looks so much like an onion. **Common Name:** Autumn crocus. **Botanical Name:** Colchicum autumnale. **Toxicity:** All parts contain the alkaloid colchicine, which has the ability to stop spindle formation in the metaphase of mitotic cell division and is used in medicine and research for this purpose. **Signs & Symptoms:** Gastrointestinal irritation, nausea, vomiting, diarrhea, bloody stool, cardiovascular collapse, convulsions and death. **Treatment:** OBTAIN MEDICAL ASSISTANCE IMMEDIATELY.

MONKSHOOD is a perennial growing from 1/2 to 1 1/2 m. (2 to 4 ft.) high, with dark green, deeply palmated leaves. The flowers are dark blue-purple to white; the shape of the sepals remind one of the hood or cowl, worn by monks, thus it's name. The petals form the lower boat- like cup underneath the cowl, making the flower unusual and striking. It has gone wild over most of North America. **Common Name:** Monkshood, wolfsbane. **Botanical Name:** Aconitum and related species. **Toxicity:** All parts, contains aconitine and others. **Signs & Symptoms:** Intense vomiting and diarrhea. **Treatment:** OBTAIN MEDICAL ASSISTANCE IMMEDIATELY.

Autumn crocus

Monkshood

ANGEL'S TRUMPET is a large, branching, shrubby annual with enormous, 25 cms. (1 ft.) long, white trumpet flowers, which are often grown in tubs for patio decoration. Unfortunately, that is where children may be tempted to pick the flowers and try to blow through them like a trumpet or, on occasion, ingestion of the seeds may occur. The seeds are encased in an ovoid, spiny capsule about 4 cms. long. There are species with smaller flowers, some of them blueish in color, but all species are poisonous. **Common Name:** Angel's trumpet, thornapple, jimsonweed, Jamestown weed. **Botanical Name:** Datura stramonium and related species. **Toxicity:** All parts contain the alkaloids hysoscyamine, atropine and hyoscine. **Signs & Symptoms:** Dilated pupils, dry mouth, elevated body temperature, intense thirst, headache, hallucinations, delirium, nausea, rapid pulse, coma and death. **Treatment:** OBTAIN MEDICAL ASSISTANCE IMMEDIATELY.

The most common shrub of the **DAPHNE** family is a small, branching plant 30 cms. to 1 m. (15 to 40 ins.) tall that flowers in early spring, long before the dark green leaves appear, and sometimes before the snow has melted. The tiny purple or pink flowers are arranged in clusters of two to five on the branches of the previous year's growth. They permeate the air with a lovely perfume, heralding spring yet to come. Later in the summer, small green berries appear, eventually turning bright red before dropping to the ground. I have found a young robin lying under the bush, dead, with a berry still in it's beak. The mother bird went on eating the berries, but discarding the seeds. **Common Name:** Daphne, spurge laurel. **Botanical Name:** Daphne mezereum. **Toxicity:** All parts, but particularly the berries, contain a glycosidic alkaloid which contains dihydroxycoumarin. **Signs & Symptoms:** Burning or ulceration in the throat and stomach, vomiting, internal bleeding, bloody diarrhea, weakness, convulsions, kidney damage, coma and death. **Treatment:** OBTAIN MEDICAL ASSISTANCE IMMEDIATELY.

Angel's trumpet

Daphne

SNOWDROP is a hardy, small, early flowering bulb, native to southern Europe. Blunt, flat grey-green leaves push through the ground very early in the spring, hiding the buds that are waiting inside to bloom. The flowers have three white, pointed sepals that open wide like a star, leaving the green-tipped, white, bell-like petals free to hang inside. When they first open, the flowers are only 5 to 7 1/2 cms. (2 to 3 in.) above the ground but as the plant matures, the length of the stem increases until the blooms fade. In Europe during the first world war, when food was scarce, some of these plants were dug and fed to farm animals, with fatal results. This is a reminder not to eat or feed to other creatures anything that is not known to be harmless. **Common Name:** Snowdrop. **Botanical Name:** Galanthus nivalis. **Toxicity:** All parts of the plant, particularly the bulb, contain alkaloids. **Signs & Symptoms:** If ingested, nausea, vomiting and diarrhea. Skin contact with the sap may cause an itchy, allergic response. **Treatment:** OBTAIN MEDICAL ASSISTANCE.

The **DAFFODIL** looks much like the snowdrop when first seen breaking through the ground, but the mature plant is more robust with larger leaves growing to a height of 15 to 30 cms., three to four times as tall as the snowdrop. The gray-green, lanceolate leaves appear early in the spring, thrusting tall above the ground, waiting for the flower buds that appear some weeks later. These stems carry one to several buds, depending on the species. The flowers may be any colour from white to deep yellow and some varieties even have pale pink petals. All members of the daffodil family should be treated with care because the sap can cause hypersensitivity reactions in some individuals, resulting in an itchy rash. **Common Name:** Daffodil, Lent lily, jonquil. **Botanical Name:** Narcissus species. **Toxicity:** All parts, particularly the bulb, contains alkaloids, including lycorine and lycoramine. **Signs & Symptoms:** Nausea, vomiting, diarrhea and sometimes, convulsions and death. **Treatment:** OBTAIN MEDICAL ASSISTANCE.

Snowdrop

Daffodil

ENGLISH IVY is a hardy, woody, climbing or creeping vine which has a habit of crawling over everything in it's path. The weight of this vine can pull down a wall or roof, if given time without pruning. Ivies have evergreen, leathery, alternate, simple, triangular leaves and berry-like seeds. The short, finger-like aerial roots attach the stems to walls and trees to heights of as much as 30 m. (90 ft.) Smaller varieties are used as pot plants and as hanging house plants. The small, yellow-green flowers are arranged in umbels and produce bitter black berries which ripen during the following year, and which, like all tissues of the plant, are poisonous if eaten in quantity. **Common Name:** English ivy, ivy. **Botanical Name:** Hedera helix and related species. **Toxicity:** All parts contain the saponic glycoside hederagenin. Usually, appreciable quantities of the plant must be ingested to cause poisoning. **Signs & Symptoms:** Excitement, difficult breathing and eventual coma. **Treatment:** OBTAIN MEDICAL ASSISTANCE.

The **HYACINTH** has a sweet scented flower and its bulbs are often forced for indoor, winter decoration. Clumps of these bulbs are used in many gardens for spring color and perfume. The individual flower is a small bell but the whole raceme makes a lovely show, be the color white, pink, blue or purple. The reddish skinned, dry bulbs can be mistaken for red onions as they lie in bins waiting to be sold in September. There have been many cases of people making the mistake of buying a hyacinth bulb in place of onions and using them in salads, with tragic results. Cooking will reduce the toxicity somewhat but they should not be eaten at all. **Common Name:** Hyacinth. **Botanical Name:** Hyacinthus orientalis. **Toxicity:** All parts, particularly the bulb, contain alkaloids. **Signs & Symptoms:** Nausea, vomiting and, diarrhea. **Treatment:** OBTAIN MEDICAL ASSISTANCE.

English ivy

Hyacinth

STAR OF BETHLEHEM, which originated in Europe, has a very small, onion-like bulb and is occasionally cultivated as a garden plant. In some places in North America, it has escaped into grasslands from old forgotten gardens. The plant is an herbaceous perennial with dark green, strap-like leaves which have a light stripe down the centre, or midrib. The thin stem grows to a height of 30 cms. (12 in.) and terminates in a number of white star-like flowers with green centres. There are reports of children being poisoned by eating the flowers and the bulbs. **Common Name:** Star of Bethlehem. **Botanical Name:** Ornithogalum umbellatum. **Toxicity:** All parts, including the bulb, contain incompletely identified alkaloids. **Signs & Symptoms:** Nausea, intestinal disorders. **Treatment:** OBTAIN MEDICAL ASSISTANCE.

LILY OF THE VALLEY is grown in cool, shady places and awkward corners that need a quick growing ground cover. The pretty green leaves, up to 30 cms. (12 in.) long, make an attractive pattern all summer, after the racemes of small, white, nodding, bell-shaped flowers have faded. They are followed by a few round, green berry-like seeds that turn red later in the season. These are attractive to children and have the potential to cause poisoning. **Common Name:** Lily of the valley, our lady's tears. **Botanical Name:** Convallaria majalis. **Toxicity:** All parts contain cardiac glycosides, including convallarin and convallamarin. **Signs & Symptoms:** Headache, nausea, vomiting, diarrhea, blurred vision, slow or irregular pulse, hypotension, cardiac irregularities and death. **Treatment:** OBTAIN MEDICAL ASSISTANCE IMMEDIATELY.

Star of Bethlehem

Lily of the valley

IRIS of many different species may be found in gardens. All species contain an irritant substance in the sap in all parts of the plant. The iris has sharp, pointed, parallel-veined, greyish-green leaves which grow to a length of 40 cms. (18 in.) tall while the flower spike can be from 30 to 75 cms. (15 to 30 in.) high. The most common garden varieties have blue to purple or white flowers and make a very showy border plant in June. **Common Name:** Iris, flag. **Botanical Name:** Iridaceae species. **Toxicity:** Irritant compound in leaves and root stalks is possibly a glycoside. **Signs & Symptoms:** Violent diarrhea, nausea and gastrointestinal irritation. **Treatment:** OBTAIN MEDICAL ASSISTANCE IMMEDIATELY.

The **CHRISTMAS ROSE** is a low growing, evergreen perennial with dark green, palmate leaves that grow on smooth stems, 20 to 30 cms. (8 to 12 in.) tall, rising from a compact, thick root system. The chalice or saucer-shaped, white, rose, or green flowers rise directly from the root. These flowers bloom, depending on the color or species, from early autumn throughout the winter, and often into early spring. Considered poisonous from the time of ancient Greece to the present day, but there are actually few documented cases. **Common Name:** Christmas rose, Lenten rose. **Botanical Name:** Helleborus niger and related species. **Toxicity:** All parts contain purgative glycosides and alkaloids, including helleborin. **Signs & Symptoms:** Severe gastrointestinal irritation, nausea, vomiting and diarrhea. **Treatment:** OBTAIN MEDICAL ASSISTANCE.

Iris

Christmas rose

LARKSPUR is a short-lived annual or perennial plant that grows as much as 2 to 3 1/2 m. (6 to 10 ft.) tall, and is often used as a background for garden plantings. The leaves are finely palmated and each flower can be over 2 1/2 cms. (1 in.) across, with the sepals elongated into spurs at the back of each flower. The number of flowers carried on each stem can be very numerous, and the colors range from pure white through all shades of blue and pink. **Common Name:** Larkspur, delphinium. **Botanical Name:** Delphinium species. **Toxicity:** All parts contain diterpinoid alkaloids such as delphinine. **Signs & Symptoms:** Gastroenteritis, nausea, constipation, respiratory paralysis and death. **Treatment:** OBTAIN MEDICAL ASSISTANCE IMMEDIATELY.

The domestic **LUPIN** that we grow in our gardens originally derived from the wild species and has since been bred into the lovely blue, red, pink and white, 1/2 to 1 m. (2 to 3 ft.) tall flower spikes seen in our gardens today. The foot long (30 cms.) stems with their unusual, hairy, palmate, grey-green leaves make a lovely foil for the pea-like flowers on their tall stalks. The seeds of this plant look like small peas and are attractive to small children. It only takes three or four of them to make a child very ill. **Common Name:** Lupin, lupine. **Botanical Name:** Lupinus species. **Toxicity:** All parts, particularly the seeds, contain the lupine alkaloids such as lupinine. Not all the lupine alkaloids are poisonous; minor changes in the chemical structure appear to cause major changes in toxicity. **Signs & Symptoms:** Labored breathing, in some instances spasms may occur, coma and death from respiratory paralysis. **Treatment:** OBTAIN MEDICAL ASSISTANCE IMMEDIATELY.

Larkspur

Lupin

FOXGLOVE is a biennial herb from Europe, grown in gardens for centuries as an herbal medicine and decorative flower. The rosette of rough, lanceolate leaves lies close to the ground in it's first year, while in the second year, the flower stalk rises up as tall as 2 m. (6 ft.) The raceme of purplish pink to white, finger-like, bell-shaped flowers hang from the side of the stalk facing the sun. Some flowers are speckled inside the bell but others have almost no spots. The foxglove has escaped into the wild in many places from British Columbia south to California, has become naturalized, and like the yellow water iris, is now considered one of our wild plants. **Common Name:** Foxglove, deadmen's or witch's bells. **Botanical Name:** Digitalis purpurea. **Toxicity:** All parts contain about one dozen steroid glycosides, including digoxin and digitoxin which have two types of effect on the heart, strengthening the force of contraction and slowing the rate of the heartbeat. **Signs & Symptoms**: Headache, nausea, vomiting, diarrhea, blurred vision, delirium, slow and irregular pulse, coma and death. **Treatment:** OBTAIN MEDICAL ASSISTANCE IMMEDIATELY.

BLEEDING HEART leaves are many-lobed or pinnately dissected, blue-green in color and grow on a red stem, 1/2 to 1 m. (2 to 3 ft.) tall. The large, fleshy roots form into big clumps that can fill awkward corners in a garden, the lovely foliage of this perennial giving cover until late summer. The graceful, drooping stems of small white and pink, heart-shaped flowers, arranged in racemes or panicles, last for a relatively short period of time. **Common Name:** Bleeding heart, Dutchman's breeches. **Botanical Name:** Dicentra species. **Toxicity:** All parts contain isoquinoline alkaloids, including aporphine, protopine and protoberberine. **Signs & Symptoms:** Trembling, staggering, salivation, projectile vomiting and convulsions. **Treatment:** OBTAIN MEDICAL ASSISTANCE.

Foxglove

Bleeding heart

CASTOR BEAN is a cultivated, shrub-like annual that grows 1 1/2 to 3 m. (4 to 10 ft.) tall with many branches, each carrying large, long stemmed, simple, alternate, palmated leaves. When mature, this spectacular plant is mostly red-purple with a little green showing in places. The fruits are oval and covered with fleshy spines. When dry, they split open and release the black and white, striped, bean-like seeds. The seeds are attractive to children, and ingestion of only three seeds may be fatal. Castor oil is extracted from the seeds following cooking, which destroys most of the toxins. **Common Name:** Castor bean, castor oil plant. **Botanical Name:** Ricinus communis. **Toxicity:** All parts, especially the seeds, contain the glycoprotein ricin. **Signs & Symptoms:** Burning of the mouth and throat, nausea, vomiting, gastrointestinal pain, excessive thirst, convulsions, uremia and death. Effects may be delayed one to three days following ingestion. **Treatment:** OBTAIN MEDICAL ASSISTANCE IMMEDIATELY.

CHINESE LANTERN has a fast growing root system with 60 cms. to 1 m. (2 to 3 ft.) tall, branching flower stalks that bear cream-colored, bell-like flowers from the axis of the leaves. The round fruit, or seeds, form inside a dry, papery green case which turns a bright orange late in the summer or early autumn. It is for these bright colored hanging cases that the plant is grown, and which make very attractive winter decorations. It belongs to the tomato family. **Common Name:** Chinese lantern, Jerusalem cherry, ground cherry, strawberry tomato. **Botanical Name:** Physalis species. **Toxicity:** The leaves and unripe fruit are toxic. The leaves contain the steroidal, cardioactive, alkaloid solanocapsine, and the fruit contains glycoalkaloids, including solanine. **Signs & Symptoms:** Headache, abdominal pain, nausea, vomiting, diarrhea, mental and respiratory depression, loss of sensation, death. **Treatment:** OBTAIN MEDICAL ASSISTANCE IMMEDIATELY.

Castor bean

Chinese lantern

HYDRANGEA shrubs come in many sizes and shapes as well as with different colored bloom trusses. These blossom trusses can be round umbels or long racemes and the colors can be white through pink to red and also many shades of blue. The miniscule true flower is surrounded by bracts, which give them their colors, while the leaves are almost heart-shaped, dark green and crinkly. Some of these plants are very tender and will not survive the winter away from the warm coastal climate, while others are quite hardy and grow well throughout the interior of the northwest. **Common Name:** Hydrangea. **Botanical Name:** Hydrangea macrophylla and related species. **Toxicity:** Buds and leaves contain a cyanogenic glycoside, hydrangin. **Signs & Symptoms:** Gastroenteritis, nausea, vomiting, diarrhea. Symptoms of cyanide poisoning. **Treatment:** OBTAIN MEDICAL ASSISTANCE.

FIRETHORN is a showy member of the rose family, with oval, dark green leaves, long sharp thorns and trusses of small white flowers. This shrub requires pruning to make it flower and to keep it under control; usually it is trained to spread against walls or fences. Late in the season the flowers give way to bright orange berries that stay on the plant all winter. Be careful, birds don't eat these berries until they are frozen, and then not willingly. **Common Name:** Firethorn. **Botanical Name:** Pyracantha. **Toxicity:** Toxic substances unidentified. **Signs & Symptoms:** Blistered tongue, abdominal cramps. **Treatment:** OBTAIN MEDICAL ASSISTANCE.

Hydrangea

Firethorn

Who would think that our queen of fruits, the **APPLE**, could harbor poison! The seeds, in common with many stone fruits, contain a compound which releases cyanide upon hydrolysis. It is on record that one man who liked the taste of apple seed ingested half a cup of seeds at one time, with fatal consequences. The leaves and tips of apple branches should not be chewed either. **Common Name:** Apple, crabapple. **Botanical Name:** Malus species. **Toxicity:** Seeds, twigs, leaves, contain the cyanogenic glycoside amygdalin. **Signs & Symptoms:** Ingestion in quantity may result in impaired respiration, loss of the voice, spasms, coma and death. **Treatment:** OBTAIN MEDICAL ASSISTANCE.

RHUBARB is well known for the edible red or green leaf stalks which are used for pie filling. The leaves, however, are so poisonous they can be dried and made into powder to dust on, or soaked and made into a 'tea' for use as an insecticide. **Common Name:** Rhubarb. **Botanical Name:** Rheum rhaponticum. **Toxicity:** Leaves contain oxalates and possibly an unidentified toxic substance. **Signs & Symptoms:** Nausea, vomiting, diarrhea, abdominal pain, oliguria, hemorrhage and death. **Treatment:** OBTAIN MEDICAL ASSISTANCE IMMEDIATELY.

Apple

Rhubarb

In the colder parts of the interior of western North America, the **LABURNUM** becomes a twiggy shrub because of winter die-back, but along the warmer coast this member of the Leguminosae family grows into a small tree. The light green, three-sectioned leaves grow profusely and show off the long. drooping, graceful racemes of bright yellow pea-shaped flowers which hang all over the tree. Later in the season, the seeds are encased in a fruit which consists of a long, flat pod. Children have been tempted to eat these 'peas', with tragic results. **Common Name:** Laburnum, golden chains. **Botanical Name:** Laburnum anagyroides. **Toxicity:** Seeds and flowers contain the quinolizidine alkaloid, cytisine. **Signs & Symptoms:** Burning sensation in the mouth and abdomen, nausea, severe vomiting, diarrhea, irregular pulse and respiration, delirium, coma, kidney damage and death. **Treatment:** OBTAIN MEDICAL ASSISTANCE IMMEDIATELY.

BLACK LOCUST is a large ornamental tree belonging to the Leguminosae family, native to eastern and mid-western USA, and naturalized from Washington to central California. The leaves are alternate, odd-pinnately divided, with two spines at the base of each leaf stalk. The pea-shaped, white flowers hang in racemes, followed by flat pods containing kidney-shaped bean-like seeds. Children have been poisoned by eating the seeds and chewing on fresh twigs. **Common Name:** Black locust, false acacia. **Botanical Name:** Robinia pseudacacia. **Toxicity:** The inner bark, young leaves and seeds contain the phytotoxin, robin and the glycoside, robitin. **Signs & Symptoms:** Nausea, vomiting, diarrhea, abdominal pain, drowsiness, disorientation, cyanosis, coldness of arms and legs, circulatory collapse, coma and death. **Treatment:** OBTAIN MEDICAL ASSISTANCE IMMEDIATELY.

Laburnum

Black locust

POTATOES, like all members of the Solanaceae family, must be used with a certain amount of caution. The green parts of the plant, particularly the fruit, which resembles it's relative the tomato, contain a poisonous alkaloid. Potato tubers which have a green coloration should not be eaten. The potato plant can also cause skin irritation in some individuals; similar effects can occur after contact with other members of the Solanaceae family, such as the tomato and eggplant. **Common Name:** Potato. **Botanical Name:** Solanum tuberosum. **Toxicity:** All green parts of the plant, particularly the fruit, contain the extremely toxic glycoalkaloid, solanine. **Signs & Symptoms:** Abdominal pain, vomiting, diarrhea, mental and respiratory depression, shock, coma and death. **Treatment:** OBTAIN MEDICAL ASSISTANCE IMMEDIATELY.

Potato

House Plants

The **OLEANDER** is a medium-sized, ornamental, evergreen shrub introduced from the Mediterranean region and commonly cultivated in North America in temperate areas with mild winters. Oleander is also found as a house or patio plant in areas where the climate is cooler. Leaves are lanceolate and leathery. The flowers are in clumps, white or pink, double or single and have a lovely perfume. All parts of the plant are extremely poisonous and mortality in humans has been high in some countries, including Hawaii. The toxic nature of oleander has been known for centuries. Even honey made from oleander nectar is poisonous. **Common Name:** Oleander. **Botanical Name:** Nerium oleander and related species. **Toxicity:** All parts contain cardiac glycosides, including oleandroside, nerioside and oleandrin. Ingestion of a single leaf may prove fatal. **Signs & Symptoms:** EXTREMELY POISONOUS. Severe gastroenteritis, tachycardia, nausea, vomiting, bloody diarrhea, dizziness, drowsiness, coma and death. **Treatment:** OBTAIN MEDICAL ASSISTANCE IMMEDIATELY.

HOLLY is a common decoration at Christmastime, with its spiky dark green leaves and bright red berries. The berries, which are only produced by the female trees, are toxic to adults when eaten in quantity, but are particularly hazardous to young children, who are attracted to their bright color. The leaves of the European holly have been used for centuries as a tonic in Europe, and in eastern North America, the native Indians brewed a strong tea from the leaves of the American holly. Early settlers prepared a weak, harmless concoction from the caffeine-containing leaves of the American species, known as "yaupon tea", which was used as a substitute for imported tea. **Common Name:** Holly, European holly, American holly, hulver bush, hulm, many others. **Botanical Name:** Ilex aquifolium, Ilex vomitoria. **Toxicity:** Berries, contain saponic glycoside. **Signs & Symptoms:** Vomiting and diarrhea.

Oleander

Holly

DUMBCANE is a popular, large-leaved plant often used by decorators. The big, showy leaves may be all green or variegated green and white to almost all white. The stems become thickened as the plant ages, and it will grow to a height of several metres. Its common name comes from the effects of the sap on the tissues of the mouth and throat when ingested. These tissues may swell to such an extent as to render speech impossible and to constrict the throat sufficiently to prevent breathing. **Common Name:** Dumbcane. **Botanical Name:** Dieffenbachia picta, and similar species. **Toxicity:** All parts of the plant. Active component believed to be calcium oxalate or possibly a proteinaceous phytotoxin. **Signs & Symptoms:** Severe irritation of mucous membranes, swelling of the tissues of the mouth, tongue and throat. **Treatment:** OBTAIN MEDICAL ASSISTANCE IMMEDIATELY.

The **CALADIUM** has delicate arrow-shaped, paper-thin, delicately marked, 25 cm. (12 in.) long leaves and long, colored stalks held well above the roots. The toxic principle of this species is believed to be similar to that of dieffenbachia species; therefore ingestion of any parts of the plant should be avoided. **Common Name:** Caladium, angel's wings. **Botanical Name:** Caladium hortuanum and similar species. **Toxicity:** All parts of the plant. Active component believed to be calcium oxalate or possibly a proteinaceous phytotoxin. **Signs & Symptoms:** Severe irritation of mucous membranes, swelling of the tissues of the mouth, tongue and throat. **Treatment:** OBTAIN MEDICAL ASSISTANCE IMMEDIATELY.

Dumbcane

Caladium

STRING OF BEADS is a succulent originating in southwest Africa, where it has evolved to survive in very hot, dry conditions. In it's native environment, the slender vine-like stems trail among the rocks and along the ground. Bright green, spherical leaves, with a translucent band on one side, are a characteristic of this species. The seed pods look similar to pea pods. The toxic seeds are distinctive, ovoid and glossy scarlet with jet black coloration for approximately one-third of the seed length from the point of attachment to the pod. It is the bright coloration which makes the seeds attractive to children. Not to be confused with the South American species Abrus precatorius, which has some similar characteristics. The seeds of string of beads are close together while in Abrus precatorius, they are spotted and several inches apart. **Common Name:** String of beads, pea plant. **Botanical Name:** Senecio rowleyanus. **Toxicity:** All parts contain alkaloids. **Signs & Symptoms:** Abdominal pain in the region of the liver. **Treatment:** OBTAIN MEDICAL ASSISTANCE IMMEDIATELY.

JERUSALEM CHERRY or **PEPPER PLANT** is often used as a house plant, and the ripe berries are bright red, resembling small cherries. The structure of the small white flowers indicate that these plants belong to the Solanum species, the same family as the garden tomato, potato, egg plant, pepper - and the nightshades. This is not a true pepper. The ripe berries are very tempting to small children, but once eaten may cause gastrointestinal irritation, stupefaction and loss of sensation. **Common Name:** Jerusalem cherry, pepper plant. **Botanical Name:** Solanum pseudocapsicum. **Toxicity:** The main toxic agent has been identified as the alkaloid solanine. **Signs & Symptoms:** Although the plant has a long-established reputation for toxicity, it is difficult to locate specific case histories. The symptoms of solanine toxicity are drowsiness, tremor, progressive weakness or paralysis, gastrointestinal irritation, and loss of sensation. Ingestion will produce pain in the mouth followed by above symptoms, depending on the quantity ingested. **Treatment:** OBTAIN MEDICAL ASSISTANCE IMMEDIATELY.

String of beads

Jerusalem cherry

AZALEAS are native deciduous shrubs of the United States, or evergreen species introduced from India and Japan which are attractive pot plants. The rose to white colored flowers are normally double and open rather like a small rose. They grow in clusters of three or four at the tip of each branch, while the dark green, ovate leaves grow at alternate, regular intervals along the stems of the plant. The related rhododendrons are evergreen shrubs. **Common Name:** Azalea. **Botanical Name:** Rhododendron obtusum and related species. **Toxicity:** All parts of the plant contain the resinoid carbohydrate, andromedotoxin. **Signs & Symptoms:** Gastro-intestinal irritation, vomiting, ataxia, coma and death. **Treatment:** OBTAIN MEDICAL ATTENTION IMMEDIATELY.

Several species of **HYDRANGEA,** some native to the eastern United States, others to Japan, are regarded as poisonous. Hydrangeas are shrubs, and grow to about 3 m. (10 ft.) tall. They have coarsely-toothed leaves which are opposite and simple, and make an attractive background. The predominantly sterile flowers grow in dense, rounded to flat-topped clusters of about 30 cms. (12 in.) diameter. Colors vary from white through pink to blue. Hydrangeas are often used as gift plants because they are not perfumed and with a little effort may be maintained for several years and brought to repeated bloom. **Common Name:** Hydrangea, oak leaf hydrangea, snowy hydrangea. **Botanical Name:** Hydrangea macrophylla, Hydrangea querifolia, Hydrangea arborescens and related species. **Toxicity:** All parts contain cyanogenetic glycosides, including hydrangin. **Signs & Symptoms:** Gastroenteritis, nausea, vomiting and diarrhea. Symptoms of cyanide poisoning. **Treatment:** OBTAIN MEDICAL ASSISTANCE IMMEDIATELY.

Azalea

Hydrangea

POINSETTIA has dark green, pointed-lobed leaves that grow 10 to 15 cms. (4 to 6 ins.) long. The clusters of tiny flowers form at the tips of the branches and are surrounded by scarlet or white to cream-white bracts that stay on the plant for months. There is only one fatality on record that is ascribed to the poinsettia, but there are many cases of sickness and skin irritation. The sap is milky-white and acrid. **Common Name:** Poinsettia, Christmas star. **Botanical Name:** Euphorbia pulcherrima, and similar species. **Toxicity:** All parts; active component, euphorbic acid. **Signs & Symptoms:** Abdominal pain, vomiting and diarrhea, skin irritation. **Treatment:** Wash sap from skin with soap and warm water. If ingested, OBTAIN MEDICAL ASSISTANCE.

CROWN OF THORNS is a member of the Euphorbia family. It has many-branched stems with large thorns and round-tipped, narrow, glossy, dark green leaves growing between the thorns. The flowers grow in small groups at the tip of each stem and are colored in shades of white and red. The sap is milky white and acrid. **Common Name:** Crown of thorns. **Botanical Name:** Euphorbia milii. **Toxicity:** All parts; active component, euphorbic acid. **Signs & Symptoms:** Abdominal pain, vomiting, diarrhea, skin irritation. **Treatment:** Wash sap from skin with soap and warm water. If ingested, OBTAIN MEDICAL ASSISTANCE.

Poinsettia

Crown of thorns

Plants which fit more than one category

Arbitrary decisions were made as to in which section some plants were placed (some appear twice, for instance, as wild and garden varieties). Please note the following entries which cross sections.

Listed as WILD, are also GARDEN varieties:
Yellow flag (iris) - page 20
Bittersweet - page 26
Sweet pea - page 26
Bleeding heart - page 52

Listed as GARDEN, are also WILD varieties:
Lupin - pages 22/50
Foxglove - page 52
Star of Bethlehem - page 46

Listed as GARDEN, are also HOUSEPLANT varieties:
English Ivy - page 44
Hyancinth - page 44
Autumn Crocus - page 38
Hydrangea - pages 56/70

Listed as HOUSEPLANTS, are also garden varieties:
Holly - page 64
Jerusalem cherry - page 68

References

Agriculture Canada - Stock Poisoning Plants of Western Canada

Arnold, Robert E. (MD) - Poisonous Plants

B.C. Drug and Poison Information Program

Cormack, R.G.H. - Wild Flowers of Alberta

Doull, J., Klaasen, C.D. and Amdur, M.O. - Casarett and Doull's Toxicology - The Basic Science of Poisons.

Dreisbach, Robert H. - Handbook of Poisoning, 5th ed.

Goldfrank, L.R., Flomenbaum, N.E., Lewin, N.A., Weisman, R.S., Howland, M.A. and Hoffman, R.S. - Goldfrank's Toxicologic Emergencies, 5th ed.

Grieve, Mrs. M. - A Modern Herbal

Hardin, James W. & Arena, Jay M. (M.D.) - Human Poisoning from Native and Cultivated Plants, 2nd ed.

Hitchcock, C. Leo & Cronques, Arthur - Flora of the Pacific Northwest

Jackson, B.D. - A Glossary of Botanic Terms

Kingsbury, John M. - Deadly Harvest: A Guide to Common Poisonous Plants

Kingsbury, J.M. - Poisonous Plants of the United States and Canada

Majak, W. & Benn, M.H. in Hui, Y.H., Gorham, J.R., Murrell, K.D. and Cliver, D.O. - Foodborne Disease Handbook - Vol. 3, Ch. 5, Glycosides.

Reader's Digest Asc. Inc. - Magic and Medicine of Plants

Rodale Press - Encyclopedia of Herbs

Seymour, E.L.D. (BSA) - The New Garden Encyclopedia

Index